Artificial
Ball Lightnings

V. P. TORCHIGIN

ISBN: 1511896922
ISBN-13: 9781511896924

CONTENTS

Introduction

A survey of two century attempts to produce Ball Lightning in a laboratory with respect to understanding that Ball Lightning nature is optical one is presented. It is shown that puzzle anomalous ball-shaped formations obtained in 19 and begin of 20 centuries can be considered as the Ball Lights with small lifetime because they disappear immediately after ceasing a gas discharge. Objects with the lifetime of several seconds have been obtained in the second half of 20 century. Bearing in mind properties of the Ball Lights, anomalous properties of such objects are explained. An explanation is presented why an erosive gas discharge is favorable for production of the objects. A survey of up-to-date attempts of production of similar objects with the lifetime about 10 s is presented. Schemes of setups for production laboratory ball lightning with the great lifetime and stored energy comparable to natural ones are considered.

It would be appear that the first successful attempts to produce in laboratory objects reminding Ball Lightning (BL) have been undertaken by English physicists Arden and Constable in the end of XVIII century. They caused a powerful charge to be accumulated in a Leyden jar and observed a small fireball illumination at the point of the discharge. The small fireball illumination was about one-fourth of an inch in diameter, was red in color, had rapid motion, and terminated with a loud explosion.

The further attempts to produce BL in XIX-XX centuries are described in books of Barry [Barry 1980] and Singer [Singer 1971] in detail enough. The solving contribution to production of artificial fireballs has been made in the USSR in the end 80-th - the beginning of 90-th years. Results of these works are presented in the book "Ball Lightning in laboratory ", published in 1994 [Avramenko 1994]. Additional attempts to produce Ball Lightning have been undertaken in 21 century also.

Various researchers differently call objects studied by them: the power-intensive plasma formation, the shone independent object, shone abnormal object, exotic vacuum object, long-living autonomous object, spherical plasma formation, etc.

We will call all these objects by abnormal objects (AOs) because their properties differ from properties of all other known objects.. Since a generally accepted theory of Ball Lightning is absent till now and its physical nature is not clear, it is very hard to judge whether these AOs are a particular case of Ball lightning.

Experiments on production of artificial BL are very valuable because, unlike natural BL, parameters of objects produced in the experiments can be carefully measured and objects can be repeatedly generated. So many experimental data were collected during two-century researches that there is no necessity to carry out additional experiments on research of AO properties.

In 2002 we have put forward the theory that no plasma is responsible for the Ball Lightning existence. Ball Lightning is a pure optical phenomenon. The Ball Lightning can be imagined as the soap bubble where the soap film is replaced by the film of strongly compressed air. The conventional white light is circulating in the film in all possible directions. The film shows itself as a planar wave guide the curvature of which is different from zero. The wave guide prevents the radiation of the light in free space. In turn, the light compresses the air due to the electrostriction pressure. This combination is closer to the light rather than to the lightning. We cut the tail in the world lightning and will call this combination by Ball Light.

The energy of the light is essentially greater than that of the compressed air. In this case the behavior of the Ball Light is determined by the forces connected with the light rather than by the conventional forces connected with material particles. Said forces were considered in numerous hypotheses but all attempts to explain the mysterious behavior of the Ball Lightning failed.

On the contrary, we have succeeded to explain all features of the Ball Lightning behavior on assumption that the forces between the light and matter play a decisive role. Our theory is mentioned in Wikipedia. All intriguing puzzles of the Ball Lightning behavior are explained in a natural way.

The following puzzles of the BL behavior have been explained.

Overcoming Gravity.

Uniform horizontal movement.

Explanation bouncing.

Bypassing obstacles.

Why directions of the wind and a motion of the ball lightning can be different.

Why ball lightning seems cold.

Explanation of circles inside perimeter.

Explanation of the ball lightning motion in a room near a floor rather than near ceiling.

Explanation how the ball lightning finds out splits, holes, and chimneys to penetrate through them.

Explanation of penetration in rooms through small splits and holes.

How the ball lightning enters the room through the window panes.

Behavior of the ball lightning near metal objects.

Why the ball lightning whistles and causes radio interference.

Why the ball lightning of large diameter takes the form of a flying saucer.

Why ball lightnings may have different colors.

Features of the disappearance of ball lightning.

How the ball lightning is catching up a flying aircraft.

It turns out that the anomalous properties of AO can be explained on assumption that there is the light circulating along AO surface in all possible directions. A reason of arising of the circulating light as well as the means that provide the circulation of light and prevent its radiation in free space can be various. But the circulating light provides anomalous properties in all known cases. Thus, we have explained anomalous properties of AOs and have shown that the optical theory of BL holds for AOs also. This gives additional arguments in favor of our theory.

At present there is an understanding that Ball Light presents a wide circle of objects of various sizes from a fraction of millimeters till tens meters, with various lifetime from microseconds till tens seconds, with various stored energy from a fraction of Joule till mega Joules. Besides, Ball Light shell can consist of not only compressed air but also various other gases.

On assumption that AOs are various forms of Ball Light, an explanation of anomalous AO properties is presented below. The term Ball Light will be used further for explanation of AO anomalous properties. In our opinion, this term characterizes their physical nature. Thus, a term "autonomous object" (AO) refers to a physical entity, term Ball Light refers to our model of AO.

Artificial Ball Light in the end of 19 century

Rather indicative experiments have been carried out by French physicist Plante in 1875-1890 [Barry]. He made observations with two flat parallel metal sheets separated by a thin mica sheet. A capacitor was connected across the poles of a number of storage batteries providing about 4000 V. The discharge formed as the mica sheet was pierced at a weak point. A small-incandesced globule was formed at the point of the discharge, apparently by evaporating of portions of the capacitor. The experiment lasted only a few minutes until the battery was exhausted and could no longer support the discharge. The luminous globule appeared to move about the surface of the plates in a random wandering motion. The discharge was also accompanied by a crackling noise.

Then Plante replaced metal and mica of the capacitor by damp surfaces separated by an air gap. The dump surfaces were formed by pads or filter paper disks moistened with distilled water. When the pads were connected across the capacitor and battery poles, a small ball-of-fire discharge formed between the two surfaces. The ball of fire occurred between moist areas and would not occur between dry areas. The discharge moved about the surfaces randomly but remained between damp areas. The ball-of-fire discharge would continue between the dump surfaces until the voltage source could no longer support the discharge. Plante speculated that the discharge caused the evaporation of the water, which aided in the current discharge between the two surfaces.

In a frame of our approach, this phenomenon can be explained as follows. The light sphere is formed in the discharge gap due to a merge of many small ball lights that arise owing the phenomenon of self-organizing of intensive light. The essence of this phenomenon is considered in [Torchigin 2007].The light is accumulated gradually in shells of ball lights and the light intensity in the shells is significantly greater than that in the spaces where the accumulation is absent. Characteristic crackling at the gas discharge can be explained by a sharp expansion of air in shells of small Ball Lights when they merge with the shell of the greater ball-of-fire. When intensive light from a small Ball Light passes completely in great ball-of-fire, compressed air in the shell of the small Ball Light begins to extend. Expansion of many small light Ball Lights causes a crackling noise. However, the intensity of accumulated light is not sufficient for formation of a lightguide from compressed air with small radiating losses. In this case, both resulting great Ball Light (ball of fire) and small Ball Lights that fill up the light energy in the resulting Ball Light, exist only in time of discharge. In the same time, the resulting light Ball Light is a genius Ball Light with relatively small lifetime and stored energy and its behavior in environment does not differ from that of all other Ball Lights. The Ball Light resides in the area where the refractive index is maximal.

If the filtering paper becomes dry because of evaporation of water in gas discharge at the great temperature, the resulting Ball Light moves to area where conversion of water to vapor takes place. Evaporation increases the refractive index in the field because of not only providing in this area additional substance in the form of water vapors, but also because of the temperature decreases at evaporation. This circumstance explains a casual movement of a fiery sphere between the plates covered by damp sheets of a filtering paper. The resulting Ball Light cannot remain on the same place as the filtering paper in this place becomes dry, and the refractive index in this place decreases. At the same time the density of air in the next areas where the filtering paper still damp, increases because of evaporation of moisture, and the resulting Ball Light moves to one of such areas.

Artificial Ball Light in the begin of 20 century

Gezehus [Gezehus 1900] studied a character of the 10 kV discharges of alternating current at electrodes of the various forms from various materials. Metal plates, a surface of water, water columns and damp sponges were used as electrodes. Between a copper plate and the surface of water separated by an air gap of 2-4 cm thickness, there was the very mobile discharge of constantly changing form. The fiery sphere moved back and forth on the plate with crackling. In several inches from the sphere, it was not felt any heat. In the glass bell placed above the device, vapors of nitrogen gathered.

The nature of physical processes in these experiments is the same as that in the Plante ones. The low temperature plasma appeared in the discharge gap and an intensive light radiation is generated at a passage of an alternative current through the discharge gap. Because of the self–organization of intense light radiation, a Ball Light appeared. Like Plante experiments, the light intensity in the Ball Light shell was insufficient for the production of a lightguide with small radiating losses. As a result, the Ball Light disappeared after ceasing the discharge. Like BL, the Ball Light radiates a light due to the light scattering of the intense light circulating in the Ball Light shell. Because of this, no heat is felted near the Ball Light.

Luminous spheres appeared at a negative electrode in the case if two thin metal wire connected with electrostatic machine were placed on a light sensitive emulsion of a photographic plate [Ledus 1899]. After separation of such sphere from a brightly luminous negative electrode, the end of the wire became dark but a small ball moved towards a positive electrode and slowly traversed the plate with extremely muddle way, sometimes stopping at a moment. To pass the distance of 5-10 cm between the electrodes from 1 to 4 minutes are required for the ball. When the ball achieved the positive electrode, the glow disappeared and a power supply behaved in such a way as if its poles were short-circuited. After developing the plate, the ball trajectory was seen at it. Electrical conductance was only in those places where the ball passed.

Appearance and behavior of the ball can be explained as follows. In that area where the ball is located, an erosive gas discharge takes place connected with decay and evaporation of the emulsion (for example, gelatin on which basis the emulsion is made). The refractive index is maximal in the area. When the matter from which the emulsion is prepared is vanished, the intensity of evaporation falls and the refractive index in this area decreases. Besides, the ohm resistance of this zone decreases because the emulsion becomes conductive. In this case, the voltage of the power supply is applied completely to the region between the positive electrode and the area. Passing current heats up the adjacent area located in a direction of the positive electrode. From this area evaporation begins, and the refractive index in the area increases. Thus, Ball Light shifts to this area.

Toepler has come to conclusion that BL reminds most of all a layered discharge – transitive between brush one and Volt arch [Toepler 1901]. Position of the shone area between electrodes can be changed by varying a current. Moving between electrodes the area can bypass the plates placed on a straight line between electrodes and penetrate through small apertures in these plates. The following arguments in favor of that the shone area is a Ball Light can be to presented. Ball Lights with small lifetime as special case BL possess many properties of BL. In particular, they keep their integrity, bypass obstacles and penetrate through cracks.

There are absolutely unexpected other evidences that both the nature of ball lights arising at gas discharges and the nature of BLs are the same. As was shown [Carrel 1996], at a certain mode of the gas discharge, the excess energy can be observed. As it is marked by Shoulders [Shoulders 2005], at some conditions of experiment occurrence of so-called exotic vacuum objects that are Ball Lights in fact are observed. In this case, a transmutation of elements on the cathode is observed. As Lewis [Levis 2003] noticed, fireballs are often observed in successful experiments with low energy nuclear reactions. There are strong reasons to consider, that physical conditions inside Ball Light are favorable for realization of low energy nuclear reactions. At last, specificity of movement of a cathode spot at the vacuum discharge reminds movement of resulting Ball Light in Plante experiments. All this testifies that Ball Lights are connected with a broad enough circle of the phenomena that take place at gas discharges.

Ball Light in the middle of 20century

A little bit different situation takes place in numerous experiments in which shone spheres exist after termination of a gas discharge. As a rule, shone spheres appeared after passing a large pulse current through a discharge gap. In this case, there is a possibility to accumulate a lot of light in a Ball Light shell. In this case radiating losses in the shell decrease and Ball Light lifetime increases. Apparently, the mentioned pioneer experiences executed by Arden and Constable at the end of XVIII century concern to such experiments

In Van-Marum experiments [Marum 1800] mobile spheres from iron, copper and lead wires repeatedly appeared. The spheres jumped aside from a cold surface. At once, we should notice that the cold surface is characteristic that the gradient of a refractive index is directed to its plane. Ball Light can jump aside from a cold surface if they initially move in a direction of this surface. Apparently, the cold surface is mentioned knowingly. The reasons responsible for the Ball Light jumps aside from a surface are considered below in the analysis of experiences carried out in the end of XX century.

The casual electrical discharge in a copper wire in the student's laboratory has created the shone sphere similar to Van-Marum one. This sphere appeared slowly slid on a table until it disappeared [Jones 1910]. On the trajectory of movement of the sphere, there was a trace of the singed spots that came to a crack in a table of 1-2 mm wide. The copper ball of 1 mm diameter has been found in a box directly under a crack. It is informed also in work [Emelin 1997] about observation of similar balls. The analysis of their appearance is presented later.

Nauer's artificial Ball Light

Rather indicative for confirmation of our approach experiments have been carried out by Nauer. His equipment is shown in Fig. 1 [Barry 1980]. In a small glass tube 1 with internal diameter of 5-7 cm the metal electrode 4 was located. The size of an aperture 2 at the end of a tube could vary. The second electrode 3 was located near the glass tube. Both electrodes settled down in the glass chamber that could be pumped out. Nauer observed that in a darkened room just after a spark occurred between the two electrodes, a luminous object appeared to rise from the upper month of the small glass tube. The luminous cloud could last for several seconds.

Nauer investigated the illuminated cloud phenomenon and found different results as the air pressure in the vessel was varied. At low pressures near 0.02 atmospheres large bright clouds formed; at higher pressures near 1 atmosphere smaller and darker clouds formed. Different gases were used: air, oxygen, propane, an air-chlorine mixture, an air-benzene mixture, and argon. Colors, intensity, and cloud dimensions varied.

It is important to note that the luminous cloud could be formed in dry air at normal atmospheric pressure. In order for the cloud to form well, the hole 2 at the end of the small glass tube 1 which enclosed electrode 4 had to be quite small. The luminous cloud produced in the normal pressure air was also visible in an undarkened room, although specific details of its form could only be seen in the dark.

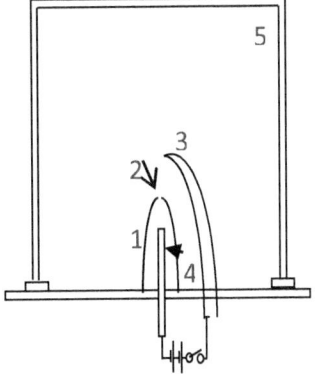

Fig.1 Scheme of the experimental setup used by Nauer where 1 is the glass tube with a small hole 2 on its face; 3 and 4 are electrodes, 5 is the vacuum chamber

It was noted that the luminous cloud rose upwards, apparently as a result of the heated gas from the spark discharge. A temperature increase of 15° C above the room temperature was measured by a thermometer in contact with the luminous cloud. The constriction at the end of the glass tube enclosing the electrode 4 improved the formation of the luminous cloud. A small hole was required at atmospheric pressure in order to produce the phenomenon. Nauer dismissed the possibility that the luminous clouds were oxidation or burning phenomena since such forms could be produced in inert gases such as argon.

We can explain simply enough the beneficial influence of a small hole 2 in the glass tube 1 on conditions of occurrence of a cloud. For successful Ball Light formation, it is necessary to increase the time during which the Ball Light is in a charge gap. All this time the Ball Light accumulates light energy in its shell. One of the most often used ways of performance of this condition is use of erosive discharge that is accompanied by evaporation of any substance from electrodes. The evaporation increases the refractive index in the discharge gap up to such degree, that its maximum is located in the discharge gap in spite of an increase of the temperature in the discharge gap.

Since Ball Light tends to move to the area with the maximal refractive index, the Ball Light remains in the discharge gap and an accumulation of light in its shell takes place. This way has been found out experimentally long ago. However, the reasonable explanation of favorable action of an erosive gas discharge on the formation of autonomous objects does not exist till now.

One more way that allows Ball Light to be in a discharge gap for a long time was experimentally discovered by Nauer. Actually, in Nauer installation a discharge gap occurs in the closed chamber. Only a small amount of gas leaves the discharge gap penetrating through the small hole. In this case, despite an increase in the temperature in the discharge gap, the density of gas in it remains almost constant and, hence, the gradient of the refractive index directed from the discharge gap is minimal.

After a termination of the discharge, the erosion ceases. As a result, the gradient of the refractive index on an axis of small hole 2 in glass tube 1 is directed outside as a part of gas has left the tube through the hole. The Ball Light penetrates through the hole in the same way as BL penetrates through a narrow crack. Nauer has experimentally shown that narrowing of the hole improves conditions of formation of a shone cloud. The same conclusion follows from the presented explanation on the basis of the concept of Ball Light. More recently the experiences with production of autonomous objects in the closed space have been repeated by Emelin [Emelin 1997] (see below).

Rather indicative for confirmation of the concept of existence of Ball Light are Nauer experiments connected with burning. Casually it was revealed, that the small traces of the benzene that has remained in the glass chamber after its clearing occurs significant influence on the appearance of shone clouds. Experiments were carried out in the closed chamber of 6-8 cm diameter and 50-200 cm long. Between two electrodes placed within the chamber, the electric spark was lit. Shone clouds appeared after the termination of the spark discharge and existed within several seconds. These clouds did not fill whole volume of the chamber, and had diameter 2-3 cm. Formation of shone clouds became more probable when hydrocarbon was thoroughly mixes in the enclosed volume of air.

We would like to analyze one very interesting Nauer experiment that enables us to discover the perfect new physical effect connected with a new type of optical nonlinearity. Studying appearance of luminous clouds at hydrocarbon concentrations smaller than that of inflammation, the following unexplained fact has been found out. The spark discharge gave no visible results in the air-propane mixture if the concentration of the propane was smaller a threshold of inflammation (about 2.8%). But at concentrations of the propane in a range from 1.4% to 1.8% yellow-green fire ball appeared of several centimeters in diameter. The ball was bright enough and promptly moved in the vessel in about two seconds. After then it disappeared *without noise*.

The energy dissipated by the spark is somewhat ambiguous in the original article. Assuming that the previous experiments were similar to those using mixtures of hydrocarbon gases – that is, a 1 ms shark carried several amperes at 1000 V – then only about 10 J may have been deposited in the gas mixture within the enclosed vessel.

All attempts to explain this paradox were unsuccessful. Using Ball Light approach the paradox is explained very simply under the assumption that the refractive index in a Ball Light shell may be increased not only by compression of air but also by drowning in the shell molecules from a gas mixture with a maximal refractive index. The refractive index of propane differs from 1 by Δn where $\Delta n = 0.00054$. On the other hand, $\Delta n = 0.00027$ for the refractive index of conventional air. It means that a double increase in Δn in the shell of a ball light can be reached by drowning molecules of propane in the Ball Light shell. For this purpose significantly smaller intensity of light is required, than that for compression of air by 2 times due to electrostriction effect.

Thus, the small impurity of propane is favorable for the formation of ball lights. The volume occupied by the Ball Light shell is smaller by several orders of magnitude than that of surrounding gas. Because of this, a small concentration of the propane is sufficient for the formation of the shell from propane. Note, that at the great concentration of propane the considered effect disappeared, as the refractive index increases not only in the Ball Light shell, but also in the space surrounding it, and the difference between them decreases.

For example, if the concentration of propane is equal 50 %, then the refractive index of the mixture is equal to $1+\Delta n$ where $\Delta n = 0.00040$ and the difference between the refractive index in the Ball Light shell filled completely of the propane and that of surrounding space is equal 0.00014. If the concentration of the propane is equal 0,1 %, then the refractive index of the mixture is equal $1+\Delta n$ where $\Delta n = 0.00027$, and a difference in the refractive indexes is $\Delta n = 0.00027$, that is by 2 times greater.

Nauer derived conclusion that the percentage of combustible gas required to form the luminous clouds is quite small and would appear to be inversely proportional to the molecular weight of the hydrocarbon. In this relation, it is very interesting to check whether a refractive index of a gas is proportional to its molecular weight. In the following table these values are given for gases taken from the handbook [Grigoriev 1991].

As is seen from the table, the greater molecular weight of a gas the greater, as a rule, its refractive index. In this case, Nauer conclusion may be reformulate as follows. The smaller a percentage of combustible gas required to form the luminous clouds the greater its refractive index. Indeed, in accordance with our approach the greater refractive index in a Ball Light shell, the better possibility of the shell to confine light circulating within it. Thus, a great refractive index of admixture is favorable for the production of ball lights.

GAS	$\Delta n \ 10^7$	Molecular weight
N_2	2793	28
NH_3	3750	17
air	2630	40
C_2H_5	6060	29
Br_2	11250	160
HBr	5700	81
H_2	1390	2
H_2O	2254	18
He	2753	8
N_2O	5150	44
HJ	9060	128
O_2	2531	32
Kr	2752	84
Xe	7020	131
CH_4	4410	16
Ne	2716	20
O_3	5110	48
NO	2970	30
CO	3340	28
SO_2	6600	64
SO_3	7370	80
H_2S	6190	34
CO_2	4197	44
F_2	1950	38
Cl_2	7680	70
HCl	4440	36

Correlation between Δn and molecular weight of various gases.

This fact is explained easily under the assumption that the refractive index in the shell can be increased not only by increasing gas pressure in the shell owning to the electrostriction effect, but also by drowning in the shell molecules of the component of the gas mixture with maximal refractive index due to new optical effect connected with separation of gas components within an intense light beam.

The effect that luminous clouds become brighter at the addition of benzene or propane is explained by the fact that Δn for these gases is significantly greater than that for other gases used in tests. In this case, a Ball Light shell can accumulate greater light that entails greater light scattering.

In the case where a shell is formed by gases with great Δn, disappearance of ball light occurs silently because there is no expansion of gas, and only diffusion of concentrated impurity occurs. If a shell is formed by the compressed gas, a disappearance of a Ball Light is accompanied by a clap because in this case sharp expansion of the gas compressed in the shell takes place.

At propane concentrations smaller than 1.8% the difference between refractive indexes of propane (in BL shell) and a mixture of air and propane (out Ball Light shell) is sufficient for Ball Light production. At propane concentrations smaller than 1.4% the difference decreases because an increase in the refractive index in Ball Light shell occurs insufficient since the density of propane molecules in the space near the shell is insufficient to form the shell consisting of propane molecules only.

Nauer observed that the luminous clouds were not affected by external electric and magnetic fields. The color of the luminous clouds was variable, but directly dependent upon the admixture gas; all colors were possible. The luminous cloud did not appear as a burning phenomenon, but rather as an electrical glow discharge – a soft glow *apparently emanating from the SURFACE of the glowing body*. Mixtures of color, uniform colors, streaks and stripes were also observed. The light from hydrogen was pale blue and that from benzene or propane was visible in a brightly lighted room.

The luminous clouds exhibited motion, generally rising, although both up and down motions were also observed. The speed of motion was measured as 0.33 – 10 m/s. No noise was detected from the phenomena at low mixture levels. A slight rustling noise was detected at high mixture values. The cloud dissipated predominantly without noise, but occasionally with a loud noise when the cloud contacted the lid of the container.

One specific trait stands out as being of some significance. The vessel was separated into compartments by a flat disk with a 7 mm central hole placed within the enclosure. The luminous cloud was normally extinguished when it contacted the disk. However, a number of times the luminous cloud appeared to pass through the hole and regain its previous size and shape.

Nauer concluded that there is a definite correlation between natural ball lightning and the laboratory-produced luminous clouds formed by a low-density hydrocarbon excitation mechanism. *It was proposed that ball lightning has no electrical nature at all.*

Describing these Nauer experiments, Barry concluded that they were most significant since Plante experiments, some 70 years earlier. On our side, we can add that description of properties of Nauer luminous clouds is in the same time description of the properties of our hypothetical Ball Light. Let us list properties of the clouds that can be easily explained if take into account that they are Ball Lights.

Nauer observed that the luminous clouds were not affected by external electric and magnetic fields. In fact, a Ball Light consists of light and compressed gas. Neither of them is affected by external electric and magnetic fields.

The luminous clouds radiate a soft glow apparently emanating from the *SURFACE* of the glowing body. Indeed, any Ball Light radiates a light from its surface because the light is concentrated in a shell that is a thin spherical layer. There is nothing besides the shell in the Ball Light.

It is not wonderful that the cloud can penetrate through a small hole and regain its previous size and shape. Process of penetration of a Ball Light through a small hole was considered in [Torchigin 2003].

The fact that the clouds did not occupy the whole space of the camera is explained by property of a Ball Light to preserve its area of surface constant.

Nauer conclusion that *ball lightning has no electrical nature at all* is valid completely because Ball Lights *have no electrical nature at all* also. Unfortunately, Nauer did not know that Ball Lights exist in the nature. If he wrote only three words spherical space soliton, a physical nature of ball lightning could be discovered a half century ago, and, possibly, at the present existence of Ball Light gained a general recognition.

Ball Light at the end of 20 century

Detailed studying of autonomous objects has been carried out in Russia in the beginning of 1990-th years. Results of many of these researches have been published in the monograph "Ball Lightning in laboratory" [Avramenko 1994]. Besides, the results of many researches are published in "Journal of Technical Physics"(JTP) and in "Letters to JTP". A powerful erosive gas discharge of several milliseconds of duration for AO production usually used.

Autonomous objects on the basis of the erosive gas discharge

A battery of capacitors was used as a power supply. AOs were produced which can exist within a fraction of second after the termination of the gas discharge. Electrodes in the discharge gap were covered by the substance that evaporated during the gas discharge. A water drop, [Egorov 2002], wax, polymers, cotton, flour from a tree [Bychkov 2004], various metals [Emelin 1997] were used as such substance. The intensity of background light in the discharge gap was essentially greater than that in Plante experiments.

It promoted occurrence of Ball Light shells from compressed air, vapors of metals or other gases. A lifetime of Ball Light surpasses lifetime of usual white light in a terrestrial atmosphere by approximately two orders of magnitude. It is explained by features of molecular light scattering in Ball Light shells. Unlike usual light scattering in three-dimensional space at which the scattering light disappears in free space, a significant part of the scattering light in Ball Light shell remains in the shell and continues to circulate within it, but in other directions. It results an increase in the Ball Light lifetime.

The further increase in Ball Light lifetime can be obtained by a substantial growth of intensity of light in a discharge gap. Such physical conditions take place at strikes of usual natural linear lightnings. The air pressure in a shell increases in such degree that the molecules within air become packed closely. It leads to substantial growth of Ball Light lifetime because fluctuations of the air density that are responsible for the molecular light scattering in the shell disappear almost completely [Torchigin 2003 Doclady Physics].

Experiments presented in [Klimov 1994] have shown that AOs appeared at gas discharge have the following properties. Their density is comparable with a density of atmospheric air; they have the low gas temperature, small intensity of light radiation, and high density of the stored energy. They are characterized by various interactions with various materials. For example, they can burn through a metal foil, but cannot penetrate through a usual leaf of a paper. They aspire to keep its integrity at a meeting with obstacles. Their lifetime is abnormal great as compared with the lifetime of an ideal plasma. Authors mark a similarity in behaviors of AO and BL.

Having known AO nature, let us consider an explanation of the selective action of the AO on obstacles. When Ball Light comes nearer to a leaf of a paper, the regions of the paper located closest to Ball Light is heated up because the absorption by the paper of the light radiated from Ball Light (sometimes tracks of carbonization are seen on a paper). The paper heats up a layer of air between Ball Light and paper owing to the phenomenon of heat conductivity. The refractive index of the hot air decreases. Since Ball Light moves along a gradient of a refractive index, Ball Light jumps aside from the paper.

Coming nearer to a metal foil, Ball Light also tries to heat up the foil by radiation of light. But the heat conductivity of the foil is greater by 2 000 times than that of the paper and the thermal capacity of the foil is greater by 15 times than that of the paper. As a result, unlike the paper, significant heating the foil does not occur. Heat spreads on the surface of the foil. It allows Ball Light to come nearer to the surface of the foil at the distance that is essential smaller than the minimal distance at which Ball Light can come to the surface of the paper.

Finally, the Ball Light would jump aside from the foil too, having come nearer to it on small enough distance and having raised temperature of a small region of the foil located closest to it. But the absolutely new mechanism at this time can come into effect. The evaporation of the foil under the action of the heat begins. This leads to the appearance of metal vapors in the area between the Ball Light and the foil. As a result, the density of gas in the space between Ball Light and the metal surface increases. This entails an increase in the refractive index in the space.

The gradient of the refractive index directed to the surface of the foil appears. Moving along this gradient, Ball Light settles down in the maximum of the refractive index that is in the immediate proximity from the foil surface. Note that the pressure of metal vapors is maximal on the foil surface as metal vapors move from the foil surface due to this pressure. However, the Ball Light moves along the gradient of the refractive index and is not blown up by the stream of metal vapors. On the contrary, the Ball Light moves in the direction opposite to the movement of the metal vapors to the area with the maximal refractive index. The area is located in the region from which the metal vapors move. The Ball Light settles down in the region where evaporation of the metal takes place and resides there all time until either it energy exhausts or it burns through the foil. In the first case, a characteristic crater is formed in the foil. Photos of similar craters are present in many works. In the second case, a characteristic hole of the correct round form in the foil is formed. When Ball Light has burned through a hole in the foil, it penetrated through the hole on the opposite side of the foil because the air on the opposite side is colder.

Experiments with AO in the discharge gap where hydrocarbon vapors from solid bodies appeared were carried out recently [Egorov 2002]. Solid bodies such as plexiglass, paraffin, wax, rosin, ground wood and their mixes were used. The scheme of the discharge chamber is shown in Fig. 2. Here a discharge occurs in the discharge gap 5 between electrodes 1 and 2. The discharge gap is formed in a plexiglass plate where a hole of 1-2 mm in diameter and 3-4 mm in length is made. A capacitor of 3,2 µF charged at voltage 300-340 V was used as the power supply. The maximal pulse current was 100-150 A. The width of the current pulse was about 1.5 ms.

AOs or in terminology of authors long-living shone formations appeared at a gas discharge. Their lifetime was about 1 s and their diameter was about 1 mm. AOs left the discharge gap approximately through 45 ms after the termination of the discharge.

Practically all observable AOs during their existence made numerous jumps on a surface of the experimental installation covered by a black paper. After decoding the video recording it was seen that AOs have almost absolutely elastic impacts. AOs, as a rule, disappeared suddenly at the next contact with a surface or in a flight.

AOs produced at the evaporation of metals were studied in St. - Petersburg [Emelin 1997]. Such AOs in the form of brightly shone yellow-red balls of 1 – 2 mm diameter appear at the erosive gas discharge with electrodes made of lead, tin or aluminum. Their falling on a horizontal firm surface is accompanied bouncing at the height of several centimeters. The number of jumps can reach up to 50.

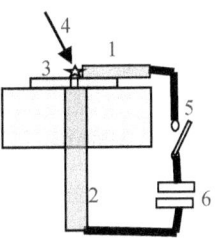

Fig. 2 Production of AO by means of the erosive discharge. Here 1 and 2 are metallic electrodes, 3 is plate for the erosion with hole 4. Capacitor 6 is discharged through switcher 5

The height of jumps gradually decreases, but the brightness of AO luminescence does not change. At a final stage of AO movement, bouncing gradually passed in sliding, then in rotating and then they stopped. There were traces in the form of the dark circles surrounded by a light deposit on the surface of the table. The circles were located on those regions from which AOs jumped aside. Traces of carbonization were seen on the surface of the wooden table in these places. The AOs luminescence usually stopped suddenly at any point of its trajectory.

According to the Ball Light concept, Ball Light shell can consist of metal vapors and polymers which refractive index is essential greater than that of the surrounding air. As follows from Grigoriev handbook [Grigoriev 1997] the value $\Delta n = n - 1$ for SO_2 is greater by $\gamma = 2,37$ than Δn for air. Analogously, $\gamma = 2,04$ for SO_3, $\gamma = 2,21$ for SH_2, $\gamma = 2,17$ for C_2H_2, $\gamma = 1,58$ for CH_4. The Ball Light has appreciable weight. Therefore, it falls on a horizontal table under the action of gravity. As Ball Light comes nearer to a surface of a table, the Ball Light heats up it, and there is the gradient of the refractive index directed upwards. It entails that a force directed upwards starts

23

to act at the Ball Light. The smaller the distance between Ball Light and the surface of the table is, the greater this force. In this respect, this force is similar to the elastic force acting on a usual ball from the surface of the table when the ball falls on its surface. As the energy reserved in Ball Light is much greater than the kinetic energy of the usual ball, the number of jumps of Ball Light is essentially greater than that of the usual ball.

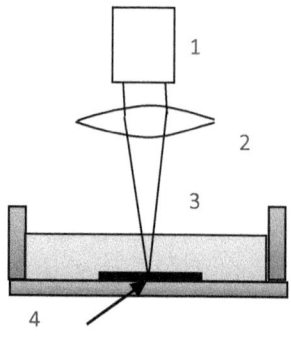

Fig. 3. Production of AOs in the liquid nitrogen 3 by means of powerful light beam from laser 1 focused by means of lens 2 on black paper 4

Carbonization of regions on the surface of the table from which the Ball Light jumped aside, confirms our assumption that, coming nearer to the surface of the table, Ball Light heats up this surface. The sudden termination of a luminescence of the Ball Light, as well as sudden disappearance Ball Light, is explained later.

Autonomous objects near 𝑙𝑖𝑞uid 𝑛itrogen

Rather impressive experiments on the interaction between AO and liquid nitrogen were carried out by Klimov et al [Klimov 1994]. Various means for AO production were used. The intense light was produced by different means, for example, by the gas discharge over the surface of the liquid nitrogen, by the impulse plasmotron, by the powerful UV flash, by the flash of the powerful flash lamp, by power pulse laser. The largest AO (diameter 10 mm) was obtained at the arc discharge directly into liquid nitrogen. An appearance of AO was observed when using electrodes made of various materials (iron, graphite, copper, aluminum, brass).

Three types of long-living AOs were observed during the experiment.

AO1 that are connected with the low intense luminescence evenly throughout the ditch liquid nitrogen and the vapor above it.

AO2 in a form of bright spherical objects diameter of 5 cm;

AO3 in a form of dark spheroids with a diameter of 1 to 2 cm.

The lifetime of AO1 is about 10-20 seconds. The life time of AO2 is 20-40 seconds. AO1 color is light blue. AO2 color changed during the evolution from bright white to deep purple. AO3 color is dirty gray. The energy stored in AO2 could reach 100-200 J. AO2 and AO3 did not appear regularly in the experiment. Ought to note that the luminescence of the liquid nitrogen was observed not only at AO production in gas discharges, but also at AO production by means of the burst of a conventional powerful flash lamp used in a photo technique. (the burst energy was about 600 J) [Dimitrov 1994].

The other most interesting results obtained in the experiments are the following.

1. Uniform blue luminescence excited in the liquid nitrogen was registered even when the UV lamp disposed near the surface of liquid nitrogen at one edge of the cell. The result is surprising, since it is assumed that the concentration of excited nitrogen molecules (and hence the luminosity of nitrogen) should decrease along the cell due to the weakening of the intensity of UV radiation (proportional to R^{-2}).

2. An anomalous excitation of liquid nitrogen was been discovered and studied for different screens and diaphragms placed in the ditch. In one experiment, the ditch with liquid nitrogen was partitioned by the tantalite opaque plate of 2 mm thickness. Nitrogen was excited by pulsed UV lamp in only one half of the ditch. Then, the partition was removed, but the sharp boundary between the excited glowing and dark unexcited parts of the nitrogen was preserved. The sharp boundary was preserved even after the forced stirring or shaking liquid nitrogen with the dielectric spatula. If there was a circle hole of 5 cm diameter in the tantalite partition, the ball of the excited nitrogen with a diameter slightly larger than the diameter of the hole was observed in the second dark half of the liquid nitrogen

3. A transparent Plexiglas plate of 2mm thickness mm was used as the partition. The liquid nitrogen in the second half was excited only in a fairly narrow band near the septum of 2 cm thickness. The form of this luminous zone resembled a fringe. There was also a long-lived intensive phosphorescence of the septum (even taken out of liquid nitrogen).

4. A glow of the liquid nitrogen decreased sharply at placing a metal object in the ditch or brought near thereto the solenoid being energized by an alternating current (frequency is 10 kHz, strength of the magnetic field is 7957.75 A/m).

5. The target in a form of the black carbon paper was placed on the bottom of the ditch for an excitation of liquid nitrogen by laser spark (see Figure 3). The target was placed in the focus of the lens. In control experiments without nitrogen, the torch was observed on the target and the target was burn. The diameter of the burned hole did not exceed 5 mm. In the case where the ditch was filled with the liquid nitrogen, a luminous spherical object of 10-20 mm diameter near the focal point of the lens was formed. The object was accompanied by a number of small secondary objects not larger than 5 mm diameter arranged randomly on the bottom of the ditch (Figure 3).

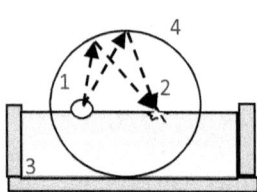

Fig. 4. Production of AOs 2 in the liquid nitrogen 3 by means of UV flash-lamp 1 located in the cylindrical reflector 4

AO moved in a usual atmosphere to a ditch with liquid nitrogen located nearby. After the termination of the gas discharge, the luminescence of the volume of the liquid nitrogen within 5 seconds was observed. The AO with precise borders of 0.5-4 mm diameter, which brightness surpasses a background luminescence of liquid nitrogen, can be seen at the bottom of the ditch, on its internal lateral walls, and also on the surface of the liquid nitrogen. The spectrum of AO radiation was in a range from 400 up to 500 nanometers. Radiation lasts during 10-30 s with gradually fading.

6. The cylindrical reflector made of aluminum or brass foil was placed around the cylindrical UV lamp as is shown in fig.4.. The lamp was placed in one of two focuses of the reflector. Light beams radiated by the lamp is concentrated in the second focus. The system was submerged in the liquid nitrogen. The spontaneous appearance of bright white AO2 of millimeter size near the second focus was observed at the moment when the lamp flashed. Local centers of

melting or even burning reflector material were detected after 5-10 flashes of the lamp. These centers looked like as craters of small diameter (about 4-5 mm) with black soot coating at their edges and the molten core of 1 mm diameter. The number of craters reached 5-7. No burning was observed in control experiments without nitrogen. The energy of the single AO2 was estimated about 6 J.

Arising from the gas discharge, AOs move in the normal air atmosphere into a nearby cuvette with liquid nitrogen. After termination of the gas discharge, the luminescence in the volume of liquid nitrogen is observed for 5 seconds. AO with clear boundaries diameter of 0.5-4 mm, which exceeds the brightness of the background luminescence of the liquid nitrogen, can be seen at the bottom of the cuvette, on its inner sidewalls and on the surface of liquid nitrogen. The radiation spectrum is in the range from 400 to 500 nm. The radiation lasts for 10-30 seconds, gradually fading. Color of AOs changes in the process of evolution from bright white to deep purple. AO in the liquid nitrogen may be prepared by means of the intense ultraviolet radiation. The flash of the powerful flash lamp (power of the capacitor bank 300 - 600 J.) also can be used for production of identical AOs. The time of AO luminescence obtained by this method is 10 - 20 seconds. A similar picture was observed in the absence of liquid nitrogen when the water vapor was introduced into the gas discharge. However, the lifetime of the AOS has been less than one second.

ᐧaᐧᐧights at ᐧacuum ᐧischarge

There are strong reasons to believe, that features of vacuum discharges can be explained by an existence of Ball Light. The vacuum in the discharge gap takes place only at the initial stage of the discharge. At a steady state vapors of the metals which have evaporated from the cathode reside in the discharge gap. As is known, the vacuum discharge can be accompanied by occurrence of avalanches of electrons, each of which contains 10^9-10^{11} electrons. Similar avalanche has been called by ecton [Mesyats 2000]. The occurrence of such avalanches can be explained as follows. Intense light in plasma from metal vapors is subjected to self-organizing. This process results in the appearance of Ball Lights shell of which consists of metal vapors [Torchigin 2004 Doclady Physics]. Interaction of AO with metal surfaces has been investigated experimentally [Avramenko 1994]. It has been shown, that the AOs are attracted to such surfaces. They can either burn a hole in such surface, or burn out a crater if the energy of AO has not sufficient.

This phenomenon can be easily explained [Torchigin 2004 Opt Comm.]. Having come nearer to the surface of the cathode, the Ball Light evaporates metal as the Ball Light heats up the surface owing to the light radiated by Ball Light. Having located in the area with the maximal density of gas, the Ball Light evaporates the metal until its energy exhausted. The evaporation of the metal and heating of the cathode are accompanied by the usual phenomenon of the thermo emission, used in radio tubes. When the energy of Ball Light becomes smaller than a threshold, Ball Light disappears. As a result, being near the surface of the cathode, the Ball Light causes the issue of an avalanche of electrons. We should notice that the character of the craters formed by Ball Light on metal surfaces at erosive gas discharges [Avramenko 1994] and the craters formed on the surface of the cathode at vacuum discharges [Mesyats 1995] are identical. The description of the casual movement of the fiery sphere in the Plante experiments reminds the description of the casual movement of the cathode spot at vacuum discharges [Luybimov 1978]. Probably, the Ball Light is responsible not only for an occurrence of ectons, but also for an occurrence of cathode spots.

Ball Lights in Ultrasound Air Stream

It is interesting enough experimental data on research of AO behavior in a non-uniform atmosphere. The scheme of experimental installation is shown in Fig. 5 [Klimov 1993]. From the working chamber 1 of volume nearby 2 cubic meters, air has been pumped out up to pressure $P_0 = 0.02-0.04$ atmosphere. After opening the electromagnetic gate 4, the atmospheric air enters the chamber through the nozzle. There was a supersonic stream 3 of 3,6 cm in diameter, moving at a speed about 470 km/s with an output of the nozzle.

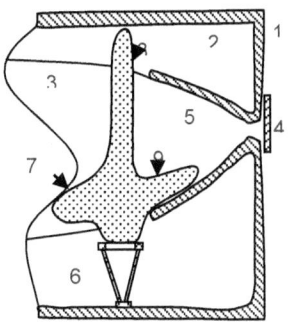

The discharge space of AO generator has a form of the truncated dielectric cone 6 with the bases of 3 and 10 mm, respectively, and 30 mm height. The internal surface of a cone was covered with wax or stearin. One of electrodes has been connected to the smaller basis; the second electrode in the form of a ring has been connected to the greater basis. The round aperture in the second electrode provided an output of AOs

Fig.5. Schematic illustration of objects flied out plasmotron 6 in supersonic stream from nozzle 5.

arising in the discharge space at the gas discharge. Speed of AO in that area where it met a supersonic stream was about 50 km/s. As follows from a photo presented in [Mishin 1992], the products arising at the gas discharge, getting in a supersonic stream, were divided into 3 shone parts 7, 8 and 9.

The part 7 consists of the exited isolated atoms that move along the stream. As one would expect, this part has a bright-white color. Authors mark "existence of such strange thing, as spreading plasmas 9 along the border of a supersonic jet upwards in a direction to the nozzle". If to admit, that the part 9 consists of Ball Light they should move along a gradient of a refractive index of that environment in which they are. In a considered case, this direction coincides with a direction of gradient of pressure of the air in a supersonic stream. This gradient is directed to the nozzle 5.

It is interesting to consider the physical nature of a part 8. This part of 2-3 m length remains strictly collimated, has bright-violet color and diameter about 1 cm. As measurements show, speed of a sound in this part is about 1 600 km/s. It is greater by 5 times than the sound speeds in air under normal conditions. As it will be shown later, sound speed increases by 5 times at compression of air, which is being at normal atmospheric pressure by 160 times. It is reached at pressure about 600 atmospheres. In this case, the density of air also increases by 160 times. Thus the refractive index of compressed air increases up to $n = 1.043$.

Consider the mechanism responsible for such great increase of pressure in the area 8. The electric current that passes through a discharge gap causes a sharp increase in the temperature in the gap. But the distribution of temperature along the radius of a cone is non-uniform. Near to the lateral surface covered by wax, the temperature is minimal, as evaporation of wax reduces temperature. As the pressure of air at various points of the gap is identical, the refractive index n is maximal near to a lateral surface of the discharge chamber.

Therefore, Ball Lights appearing in a discharge gap move in a direction to a lateral surface of the discharge chamber. Having reached a lateral surface, the Ball Lights merge with other similar Ball Lights and form the great Ball Light size of which gradually increases, until the diameter of its shell does not become equal to the diameter of a discharge chamber. Intensity of light in the shell constantly increases owing to merge to the shell new Ball Lights. It leads to increase in the electrostriction pressure and, hence, the pressure of air inside of the shell.

The following circumstance testifies that the shell receives light from small Ball Lights as a result of merges. Numerous Ball Lights of small diameter (some tens micrometers) have greater radiating losses. These losses depend on the wavelength of the light circulating in the shell. The greater wavelength is the greater radiating losses. As a result, the light radiation of a long - wave range of a spectrum, is radiated more quickly. As a result, such Ball Lights get violet color. This phenomenon explains the violet color of the great Ball Light.

The density of air and its refractive index n in a discharge gap gradually decrease, as hot air extends and leaves the discharge chamber. As a result, moving along a gradient n, the Ball Light leaves the discharge region too. The same concerns and to that part

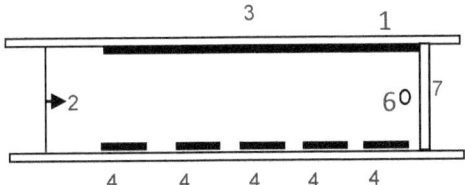

Fig. 6. Scheme of setup for measuring parameters of shock sound wave traveling through a glowing discharge

of a shell of the great Ball Light which is located near the great basis. This part starts to penetrate into the supersonic stream, but does not move with it because a part of the shell of the great Ball Light, located near the small basis remains in the discharge chamber.

Notice, that diameter of this part coincides with the diameter of the great basis. The part of the shell of the great Ball Light located near lateral walls of the discharge chamber continues to absorb small Ball Lights that continuously appear in the whole volume of the discharge gap. In this case, the great Ball Light proves as a unit with the shell where the intensive light circulates in every possible directions. Such Ball Light reminds deformed BL penetrating through a small hole. If the lifetime of such Ball Light would be great enough its form could be spherical after a termination of the discharge.

Wax in this experiment plays a triple role. First, at the evaporation of wax there are gaseous hydrocarbons. Their refractive index is greater, than that of air. Such mixture of the air and hydrocarbons is favorable for a Ball Light generation [Torchigin 2004 Opt. Comm.]. Secondly, evaporation of wax provides in the discharge gap an existence of great n for a long time as new portions of wax turn in vapors. It favorably affects the process of accumulation of light in the great Ball Light. In the third, evaporation of wax reduces temperature near lateral walls of the discharge chamber that promotes an increase in a refractive index in this area, and arising the gradient of the refractive index directed to this area. Such gradient provides movement of again arisen Ball Lights in the zone where these Ball Lights are merged into greater

Ball Light. As a result, the light intensity of the great Ball Light gradually increases.

Another rather indicative experiment on studying the properties of AOs is connected with the analysis of reflection from a flat wall of the shock sound wave propagating in small-ionized air [Mishin 1992]. The scheme of experimental installation is shown in Fig. 6. Here 1 is working section of a shock pipe with 10x10 sm2 cross-section in which the shock wave 2 at 400-600 km/s speed of a triangular structure of pressure behind its front propagates. In a working section of a shock pipe, the decaying pulse discharge with an average density of a current 30 mA/см^2 and 3 ms duration of burning was created. The discharge was created directly ahead of arrival of the shock wave.

Conventional air at the pressure 0.003-0.03 atmospheres was used as a working gas. The sizes of plasma area were made 10x10x20 cm^3. In the plasma zone at the distance of 15 cm from its beginning the flat quartz wall for reflection of the shock wave was settled down. The pressure of air was measured by means of the piezoelectric sensor 6 located on 5 cm distance from the quartz wall. As shown in [Mishin 1992], pressures of air fixed by the sensor behind the falling and reflected waves decreases at presence of the gas discharge.

For an explanation of this abnormal phenomenon, it is possible to admit, that at this time there are numerous small Ball Lights. The pressure inside of Ball Light shell is much greater than that in surrounding space. As a result, the average pressure of remaining air fixed by the sensor decreases. Like previous cases, small Ball Lights can merge in one great Ball Light. This assumption is supported by information, that "during experiments occurrence of specific plasma formation with unusual parameters near the wall in the region of hot layer was revealed". This formation occurred after reflection of a shock wave from a wall 2. Initially, this formation covered area of 3,5 cm, but then it was gradually compressed up to 2 cm ".

Ball Light in the begin of 21 century

❒a❒❒❒ghts arising at ❒va❒oration of ❒ ater

Like the previous experiment at the evaporation of wax, favorable conditions for occurrence of the great Ball Light at evaporation of water can be created. This great Ball Light represents the isolated shone sphere which diameter is approximately equal to the diameter of a vessel from which it appears. Lifetime of such Ball Light is about 0.3-0.8 s. In the experiments described in [Egorov 2004], the role of wax plays a drop of water placed on one of electrodes as shown in Fig. 7a. The gas discharge occurs between electrodes 2 and 3. The electrode 2 represents a copper ring placed on a bottom of a polyethylene vessel of 18 cm in diameter. The vessel is filled with usual water. The height of a water column is 18 cm. A structure of an electrode 3 is shown in fig. 7b. Here the quartz tube 4 with a metal or coal electrode 7 of 5-6 mm in diameter *contains about 0.1 milliliter of water 5. The tube towers* on the height from 3 up to 5 mm above a water level 6 in the vessel. A light gas in the form of vapor between the electrodes was formed upon the water evaporation induced by electric discharge of a capacitor battery that had a capacitance 600 μF and was charged up 5 KV. Egorov and Stepanov emphasize that there was an optimum potential difference between electrodes. This is likely associated with the water evaporation regime. Water is sprayed at high voltage.

Note that water in one or other forms is present in many known experiences on the artificial BL production. Water is used when forming long-living plasmoids [Egorov 2004] by means of the electric discharge. Water is also used when luminous spherical structures are formed by an electric explosion of thin wires by the intense electric current from a capacitor battery [Urutskoev 2002]. In those experiments, the electrodes had a shape that a hemispherical rather that cylindrical shock wave was formed at the initial stage. In this case, the refractive index in the cylindrical layer (CL) increases due to an increase in both the temperature and the amount of gas because of water evaporation. The velocity of propagating CL in radial directions decreases near the liquid surface. As a result, there is only one hole in CL. The hole collapses under the action of forces that arise due to the intense light. These forces compresses the gas. Similarly, a hole in a soap bubble collapses due to surface tension of forces in a soap film. In experiments [Urutskoev 2002] the

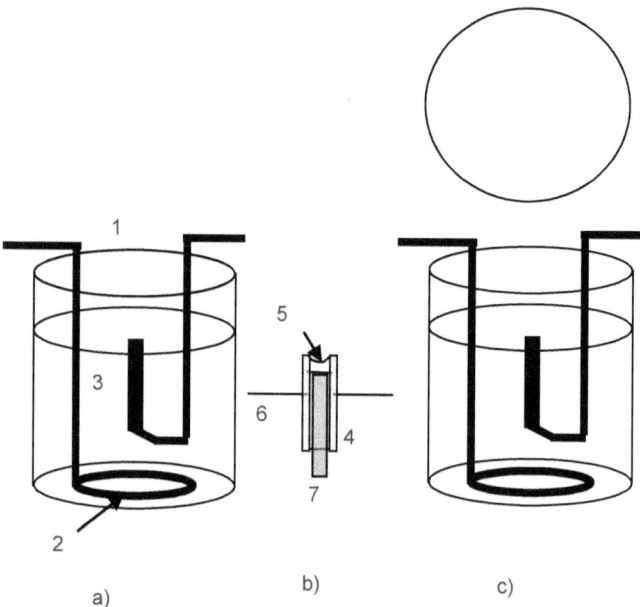

Fig. 7. Production of AOs in a vessel with water. a) location of electrodes in water; b) structure of central electrode 3

capacitor battery had energy 7.5 kJ, which is obviously insufficient. The energy of long-living BL in which intensive light compresses air in such degree, that losses of light in it considerably decrease, should be much higher. Multiple discharges can be used in the following manner. The first discharge evaporates water and prepares the CL. Then, the subsequent discharges introduce light energy into the CL. Such pulsed discharges are often observed in nature [Barry 1980].

Ball⸘ghts ⸘enetrating through ⸘rans⸘arent ⸘ a⸘

From presented consideration can be concluded, that the gas discharge in a hermetic vessel is favorable for the generation of Ball Light. Really, in this case density of gas in a vessel and, hence, a refractive index of gas can increase only at the gas discharge because of evaporation of electrodes. Experimental confirmation of this assumption is presented in [Pirozerski 2003]. The scheme of installation in which the AO are generated, is shown in fig. 8. Here 1 is a polyethylene tube with the transparent walls filled with air under normal conditions, 2 and 3 are aluminum electrodes.

To provide full tightness of the internal chamber around

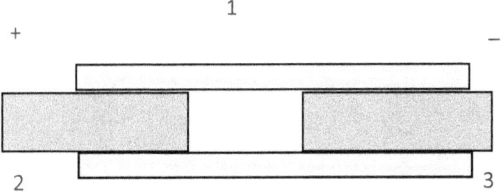

Fig 8. Scheme of discharge camera between electrodes 2 and 3 where autonomous objects penetrating through transparent wall of tube 1 have been produced

electrodes the bandage has been placed. Video recording of the process shows, that after the termination of the discharge in a tube, it got a barrel shaped form. Inside of the tube, a weak luminescence was seen that suddenly disappeared. At this moment some shone AOs in diameter about 1 mm appeared outside the tube. Their occurrence was accompanied by crackling. On the several frames of the video, it is visible, that the AOs float in the air, and then fall. AOs that fell on a surface of a table jumped several times and disappeared completely. The tube remained tight.

Meaning, that walls of the tube are impenetrable for any particles (a molecule, atoms, ions, electrons), it is necessary to recognize, that all such particles remain inside the tube. As for light, it passes through the transparent walls of the tube almost free. Process of penetration of Ball Light through transparent walls of a tube is similar to the process of penetration BL through windowpanes [Torchigin 2003].

That fact the Ball Lights leaved the tube with some delay can be explained as follows. At the gas discharge, there is an erosion of the electrodes and pressure in the tube increases. The swelling of a tube testifies to it. After the termination of the gas discharge vapors are condensed, and pressure in the tube falls. The refractive index in the tube at the great pressure is greater than the refractive index in the space outside the tube. In this case, the Ball Lights remain inside the tube. At reduction of pressure with cooling electrodes, the refractive index decreases and can become smaller than that outside the tube, as the temperature inside the tube is greater, than outside. At that moment, Ball Lights penetrate outside through the transparent walls of the tube.

Similar situation is described by Stakhanov [Stakhanov 1979] in event number 110. The luminous ball of 2-3 mm diameter has flown from the signal neon Ball Light at the moment when the Ball Light was turned on. The Ball Light is burned out. This experiment was repeated successively once more. The luminous ball of 3-4 mm penetrated through the glass balloon of the Ball Light at the moment when the Ball Light was burned out. The Ball fell on Watman sheet paper, jumped several times on it at the height of several centimeters and, disappeared without leaving a trace. However, attempts to repeat this result in further experiment were unsuccessful

Ball Light arising in the microwave oven camera

Analysis of sources in the Internet by using key words "ball lightning" shows that besides numerous evidences of eyewitnesses about observations of natural ball lightning, there are many reports about attempts to produce ball lightning in a laboratory. About half of the reports are related with successful attempts to produce AO in microwave oven. A burning candle is located in a microwave oven camera for this purpose. The height of the candle and its location are chosen in such a way that a flame of the candle is located in the place where the microwave field is maximal. After then a front door of the microwave is closed and the microwave magnetron is turned on. A luminous cloud is formed from the flame of the candle that rises upwards. Thus, the candle usually dies away, but in some experiences continues to burn. Experiences of professor Sven are rather indicative in this respect. His video film is presented on a site

http://www.youtube.com/watch?v=qoZlIUXEpns

If the candle is covered with the glass turned upside down so the bottom is a little above a flame of the candle, it is possible to observe a luminous formation which continuously changes its form for a long time (see video film on a site

http://www.youtube.com/watch?v=CsxFINLtaK8&feature=relat ed).

The glass turned upside down interferes movement of plasma upwards, and it remains near maximum of the microwave field. In this case, the luminescence of plasma is observed all time while the microwave is turned on. However, as soon as the microwave is turned off the luminescence at once vanishes.

This phenomenon is explained as follows. The microwave energy heats up and ionizes products of burning of the candle transforming them in plasma. As the temperature of plasma is greater than that of air, plasma rises upwards leaving maximum of microwave field and then disappears. However, at some positions of the candle the microwave field supports the existence of plasma for a long time. As is known, plasma is an excellent nonlinear optical medium that is favorable for Ball Light originating. But the luminous formations have not precisely outlined borders. This fact testifies to Ball Light absence. Indeed, circulating light could be responsible for precisely outlined borders. However, the light intensity at microwave discharge is small. Because of this, if even there are some Ball Lights at time of discharge they disappear immediately after turn off microwave energy. In this relation, such Ball Light reminds Ball Light obtained by Plante above 120 years ago.

Ball Light arising at the electrical discharge on the water surface

Egorov and Stepanov experiments concerned production of long-living luminous objects in a glass of water [Egorov 2002] has been repeated also. Scientists from Berlin presented a video about production of similar objects on site

http://www.youtube.com/watch?v=F2CvTNWBias .

The paper in American journal New Scientist with a description of these experiments is accessible at the address

http://www.newscientist.com/article/dn9293-physicists-create-great-balls-of-fire.html

Ball Light arising in the microwave oven camera

Analysis of sources in the Internet by using key words "ball lightning" shows that besides numerous evidences of eyewitnesses about observations of natural ball lightning, there are many reports about attempts to produce ball lightning in a laboratory. About half of the reports are related with successful attempts to produce AO in microwave oven. A burning candle is located in a microwave oven camera for this purpose. The height of the candle and its location are chosen in such a way that a flame of the candle is located in the place where the microwave field is maximal. After then a front door of the microwave is closed and the microwave magnetron is turned on. A luminous cloud is formed from the flame of the candle that rises upwards. Thus the candle usually dies away, but in some experiences continues to burn. Experiences of professor Sven are rather indicative in this respect. His video film is presented on a site

http://www.youtube.com/watch?v=qoZlIUXEpns

If the candle is covered with the glass turned upside down so the bottom is a little above a flame of the candle, it is possible to observe

a luminous formation which continuously changes its form for a long time (see video film on a site
http://www.youtube.com/watch?v=CsxFINLtaK8&feature=related).

The glass turned upside down interferes movement of plasma upwards, and it remains near maximum of the microwave field. In this case the luminescence of plasma is observed all time while the microwave is turned on. However, as soon as the microwave is turned off the luminescence at once vanishes.

This phenomenon is explained as follows. The microwave energy heats up and ionizes products of burning of the candle transforming them in plasma. As the temperature of plasma is greater than that of air, plasma rises upwards leaving maximum of microwave field and then disappears. However, at some positions of the candle the microwave field supports the existence of plasma for a long time. As is known, plasma is an excellent nonlinear optical medium that is favorable for Ball Light originating. But the luminous formations have not precisely outlined borders. This fact testifies to Ball Light absence. Indeed, circulating light could be responsible for precisely outlined borders. However, the light intensity at microwave discharge is small. Because of this, if even there are some Ball Lights at time of discharge they disappear immediately after turn off microwave energy. In this relation, such Ball Light reminds Ball Light obtained by Plante above 120 years ago.

Ball Light in a form of bouncing glowing balls

There were reports of new types of autonomous objects on the border 20 and 21st centuries. Unlike previous luminous autonomous objects (AO) whose lifetime was a fraction of a second, the lifetime of the new types reached several seconds. The

first report of such objects appeared in 1997 [Emelin 1997]. A photo of bouncing glowing balls (GBs) in a form of their

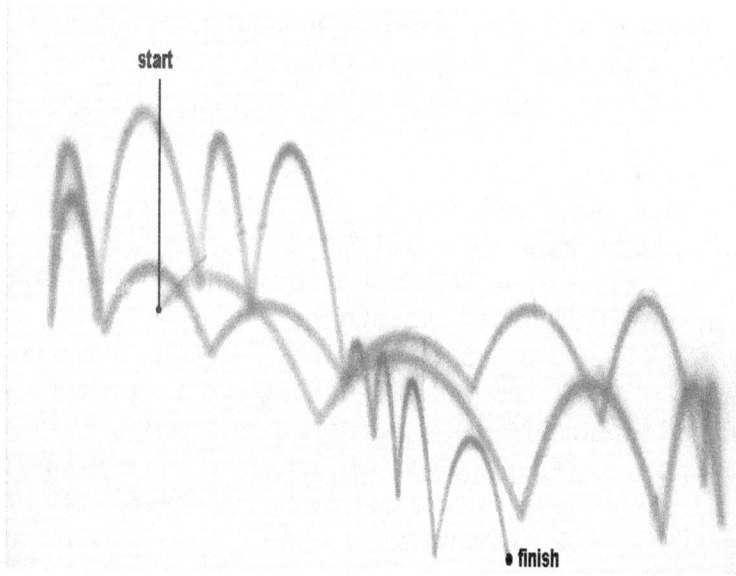

Fig.9 Trajectory of bouncing ball

continuous trajectory obtained in a dark with an open shutter of photo camera is shown in Fig. 9. Bouncing glowing balls of a few millimeters size are observed, which had abnormally high jumping ability. Balls disappeared without a trace at any point of its trajectory. GBs can make on the surface of an ordinary table about 50 jumps. Since there is a horizontal component of GB velocity in the photo, the GB trajectory is a set of clearly defined parabolas. Each parabola corresponds to one jump. There are 20 parabolas in the photo. Lifting and dropping branches of parabolas are identical in their durations and therefore the time of lifting t_{LIFT} and the time of dropping t_{DROP} are identical

For the sake of justice ought to note that an analysis of the bouncing ball in detail was presented by Stakhanov in case number120. A glowing ball about 4 mm diameter raised at short

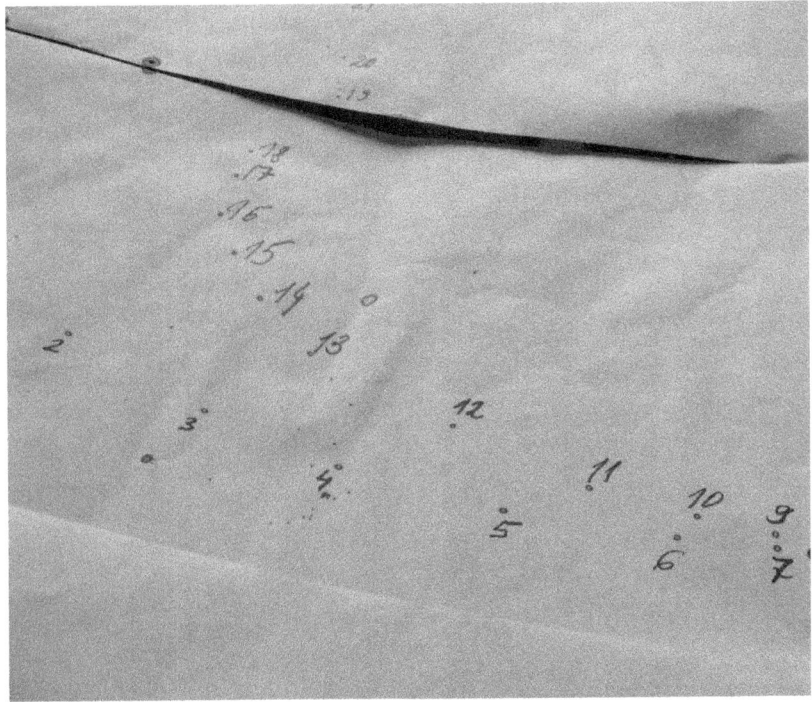

Fig. 10 Trace of the bouncing ball sheets of fax paper. Number near the spot shows a number of the jump.

circuits. The ball made 12 jumps for 4 second on the rubberized tape of the conveyor. The height of jumps was about 1.5 cm. The ball diameter was decreasing gradually till 1 mm and then the ball disappeared without leaving trace. Stakhanov calculated the time interval between adjacent jumps. It turned out that the time was greater by 3 times than the time of a solid ball in the field of gravity. Stakhanov concluded that the density of the ball is close to the density of the air.

Another group of experiences is connected with attempts to produce AOs at the powerful gas discharges. Notions about a ball lightning presented by Abrahamson and Dennis [Abrahamson 2000] are used. According to these representations, ball lightning exists owing to oxidation of silicon nano particles in an

atmosphere. Such particles are formed at the stroke of a usual linear lightning in the ground as a result of reaction in thickness of the ground of oxides of silicon and carbon. Though these

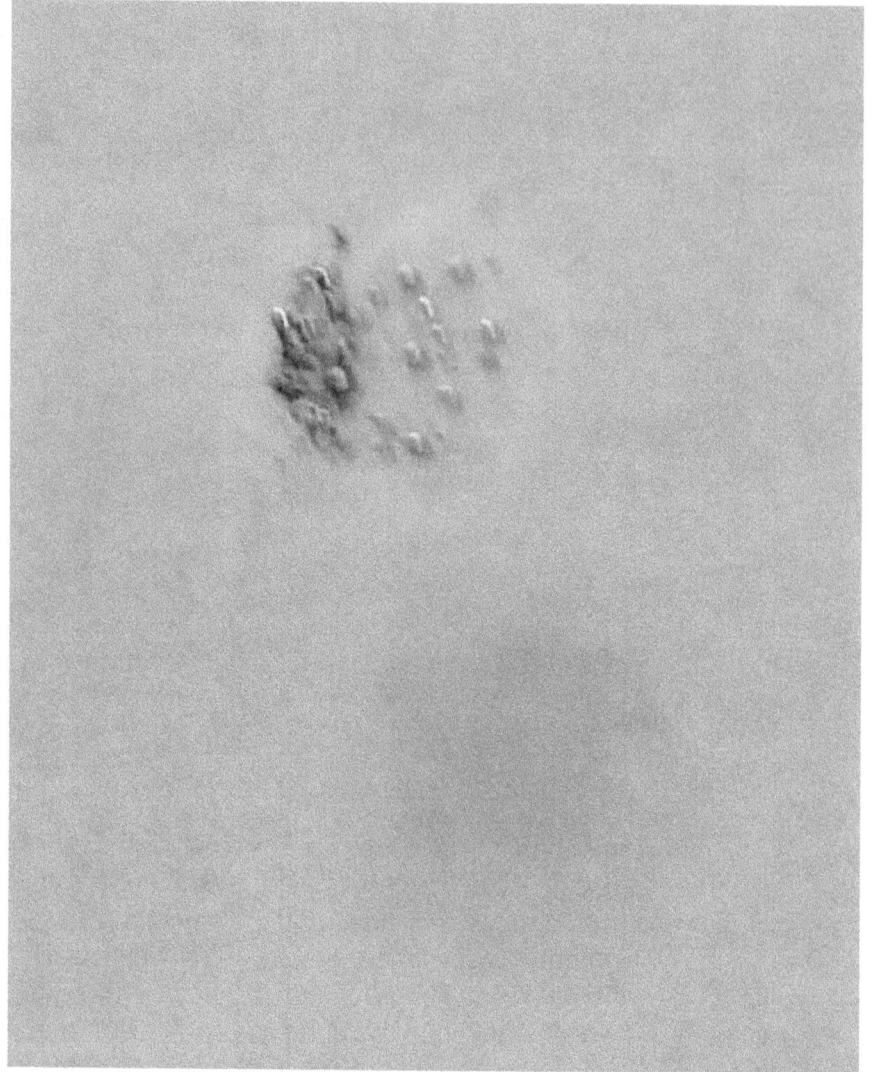

Fig. 11. Bouncing Ball that is separated at the majority of small balls at a strike against the horizontal windowpane placed on the table. The dark spot below the ball is the trace of initial heating of the surface of the table through the glass.

representatives cannot explain the facts of BL movement against a wind, penetrations through window panes, catching up airplanes, however, they promote some successes at attempts to product artificial BL. Really, as follows from these representations, an erosive gas discharge is required at which additional products evaporated from electrodes should be delivered in the BL region.

Brazil scientists reported in 2007 that glowing balls (GB) can appear at conventional arc discharge with small pieces of Si wafer introduced in the discharge gap [Paiva 2007]. Unlike conventional sparks that accompany an arc discharge, GBs have unusual properties. In first, their lifetime is essentially greater than that of the sparks and can achieve eight seconds. In second, they jump in process of their moving, can bypass obstacles, and penetrate through splits which width is smaller than their diameter. In opinion of the scientists, these GB properties remind behavior of Ball Lightning. Video film about the behavior of these objects can be seen on a site http://www.youtube.com/watch?v=fsu8IaaVsvk

These experiences have been repeated by the Holland researchers. Their video film is presented on a site http://www.youtube.com/watch?v=QLTPELhKAYM andhttp://www.youtube.com/watch?v=QLTPELhKAYM&mode=related&search .

An electronic paper about the results of their experiments entitled "Artificial Ball Lightning produced in the laboratory!?" is presented on the site
http://members.chello.nl/r.dekker49/boBall_Lightliksem/boBall Lightliksem.htm

There are 2 video films from viewing of which it is possible to be convinced that lifetime of some luminous balls is about 6-8 seconds and that they jump aside from a table more than 10 times. The height of jumps varies, and the height of the subsequent jump can be greater than that of previous one. Similar film is presented by the American researchers on a site
http://www.youtube.com/watch?v=T7fUKEGyxS8&NR=1and Spanish investigators on site
http://www.youtube.com/watch?v=IYSie5YsaOI .

Thus, now foreign researchers have reached that level which was shown by the Russian researchers 5-10 years ago. As can be seen, foreign researchers should investigate more many still that have been investigated in Russia already.

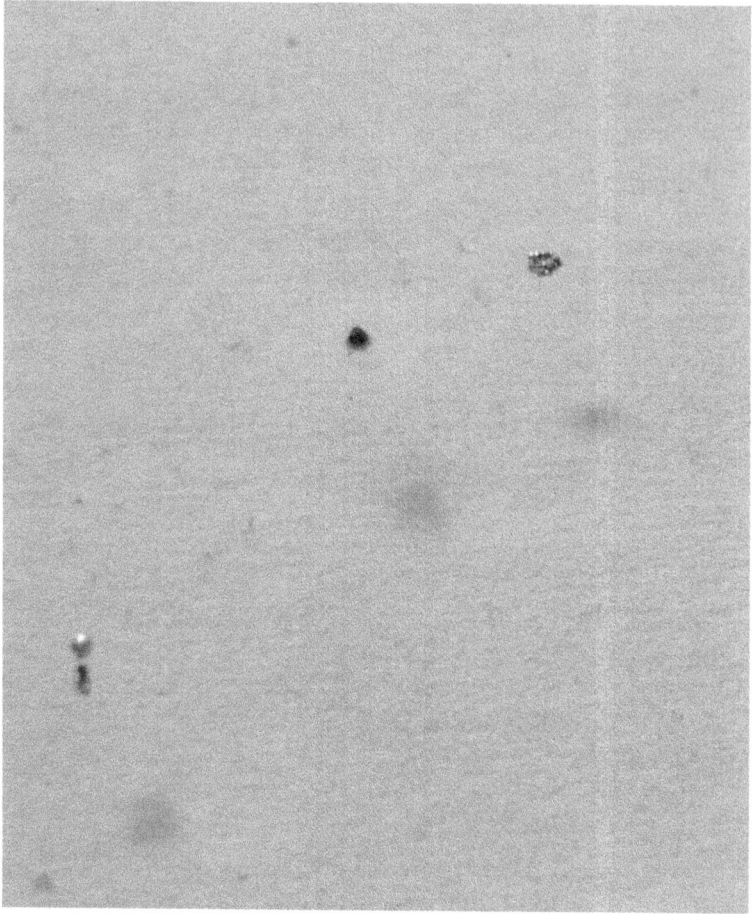

Fig. 12 Traces of balls on the place surface of glass of 4 mm thickness. As is seen, there is no bouncing.

The most informative is video by V. Gooses, P. de Graaf and R. Dekker from Holland, Eidhoven [Gooses 2009]. They accompany their video by description in detail of their experimental installation and features of experiment. They also think that their GB is a certain form of Ball Lightning and present the most known at present theory of Ball Lightning of A. Abrahamson and J. Dinniss [Abrahamson] But the question whether similar GB is a certain version of Ball Lightning remains open. It is a good chance to apply our theory to resoles this problem.

One can see in the video presented in [Gooses 2009] that the GB can jump on the surface of the white table during 8 s. The

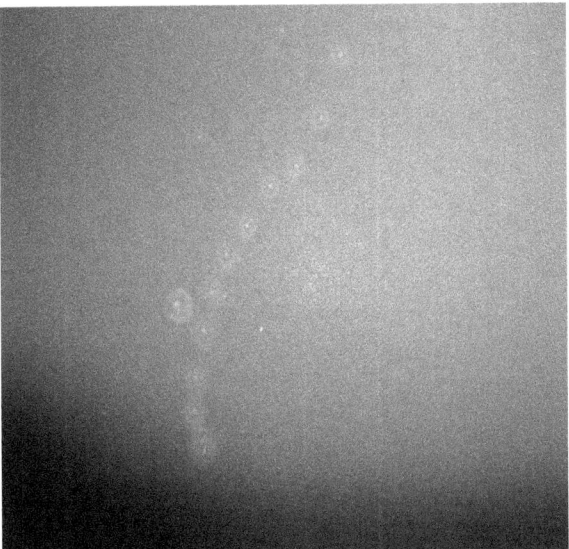

Fig. 13 Trace of the bouncing ball on the black paperboard

maximal height of the GB is about 10-20 cm and the maximal height decreases gradually with time. Nevertheless there is the jump maximal height of which is greater than that of the previous jump (second attempts, third jumps in sixth second in [Gooses 2009]).

Besides, we carried out some additional experimental investigations to clarify a physical nature of GB. Our installation is described in detail in [Torchigin 2010]. The piece of the silicon wafer of 0.4 mm thickness and 2*2 mm2 area is placed between two wolfram electrodes that are connected with the battery of automobile accumulators of 36-48-69 through switcher that capable to switch up to 300 A current. Initially the current through the silicon wafer is absent because the wafer is dielectric. The initial voltage pulse applied to the electrodes breaks down the dielectric. Then the switcher is turned on. As a result, the silicon wafer heats up and melted.

Drops of the liquid silicon in a form of GB fall on the horizontal surface of the table. The further behavior of the drops is identical to that in the mentioned above experiments.

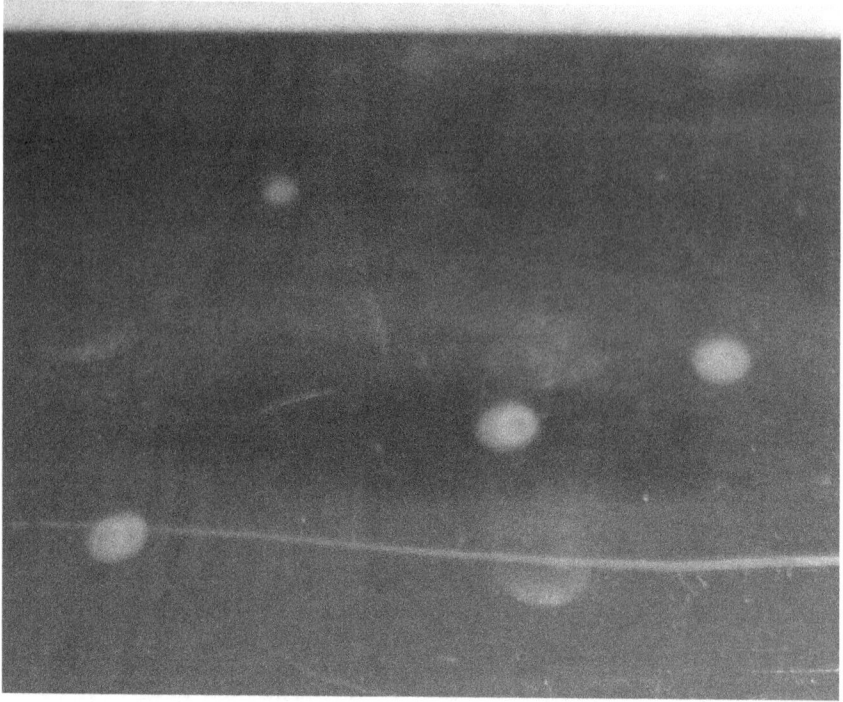

Fig. 14. Trace that the bouncing ball leaved at tantalite at bouncing. The small ball is seen in the top right-hand corner. Possibly, the light energy within the ball becomes insufficient for next jump

It turns out that the study of the character of the behavior of the drops on various surfaces can give much additional information about physical processes that takes place at the interaction between the GB and the surface. Fig. 10 shows traces of GBs at impact with the surface in a form of the thermal paper used in faxes. The fax paper has the following properties. The paper becomes dark at heating above 80^0 C. If the face of a hot circle rod heated at the temperature smaller than 250^0 C contacts with the paper surface, a trace in a form a dark circle remains on the paper. If the rod temperature is greater than 250^0 C, the trace in a form of a white circle surrounded by the dark ring remains on the paper. An

Fig. 15. Photo of the ball as compared with the ball in the ball-point pen. The surface of the ball is brilliant. It is supposed that the ball consists of the silicon and its surface is covered by silica.

appearance of the dark ring is explained by propagation of heat along radial directions with the gradual decrease of the temperature. The darkness of the ring decreases with an increase of

the distance to the center because the temperature of the paper decreases also. The area where temperature is in the range 80^0 C$<$T$<250^0$ C occurs dark.

Sequential collisions of GB with fax paper are numbered by hand in Fig.10. As is seen, there is about 20 GB jumps on the fax paper. Thus, there are no doubts that the GB heats the obstacle at collision. It is not clear in advance either a direct contact with obstacle takes place or the paper heating takes place due to light radiation at GB approaching to the obstacle.

To clear up the process of GB colliding with the surface of the obstacle the following experiment was carried out. A glass plate from windowpane of 4 mm thickness was used as the obstacle for GB. The glass plate has property to pass the radiation through itself. In other words, the windowpane is not heated by radiation

noticeably because the radiation penetrates through the glass. In this case, the GB hits the glass surface and is separated into a set of

47

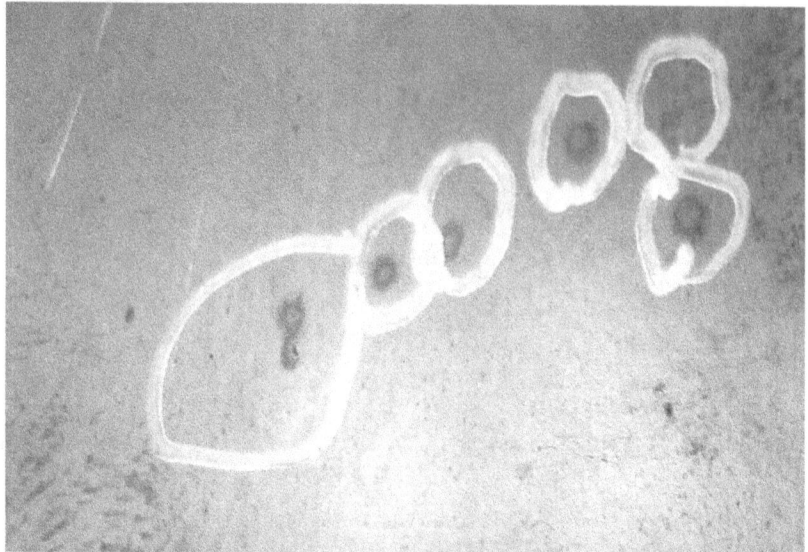

Fig. 16. Trace of the bouncing ball on the foil getinaks. Each trace is rounded by the circle made with felt-tip pen.

small balls as is shown in Fig.11. As is seen, GB damages a little a fragment of the smooth top surface of the glass plate. The dark spot is seen on the white paper under the glass plate. Bouncing GBs on the glass surface were never observed. As is seen in Fig. 12, ball are located near the traces where they collide with the glass plate.

Traces of bouncing GB on the black sheet of paper are shown in Fig. 13. One can see that there is the small white spot is surrounded by the circle white spot of greater diameter. In is explained by the fact that the GB bouncing in all experiments is accompanied by the white smoke that is produced by the GB in all time of its existence. This is valid for the Emelin *et al* experiments, [Emelin 1997]for the Paiva *et al* experiments [Paiva 2007] for the V. Gooses, P. de Graaf and R. Dekker experiments [Gooses 2008]

as well as for our experiments [Torchigin 2010]. The same white traces that the bouncing GB leaved on the tantalite plate are seen in Fig. 14. The small ball is seen in the top right-hand of the figure. This is the last contact of GB with the tantalite. As is seen in Fig.13 and Fig. 14, the distances between adjacent spots are not identical.

The ball that is seen in the top right-hand of Fig. 14 is shown in Fig. 15. One can see that the surface of the ball is brilliant.

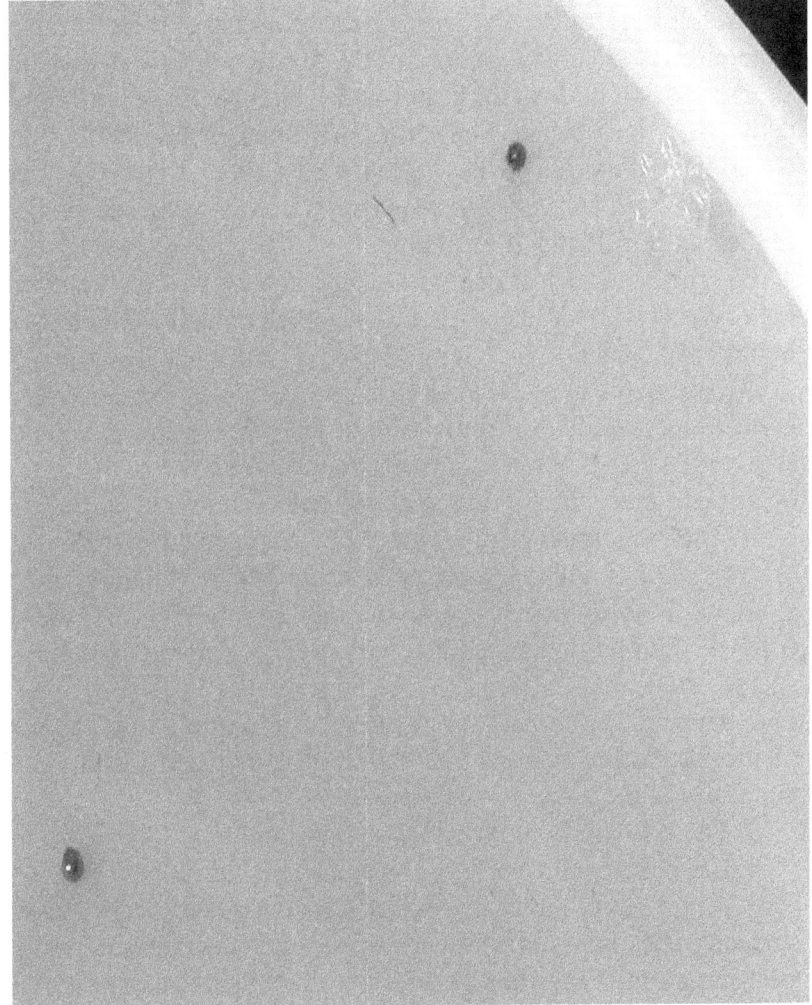

Fig. 17 Photo of the bouncing balls on the plastic vessel filled by water. The traces of burning of the bottom are seen near the balls.

A behavior of GB on the surface of thin foil bounded to getinaks is shown in Fig. 16. The traces of the GB are faintly visible. The traces are rounded by circles made with the white felt-tip pen. Unlike the glass plate, several jumps are seen.

A behavior of GB on the surface of water has been studied also. It turns out that GB slides on the surface without decreasing its brightness in several second. As a rule, the glowing ceases with GB disappearance. The events are observed when the sliding GB jumps on the surface once. Sometimes the glowing GB penetrates the water. Two GBs on the bottom of the plastic vessel is shown in Fig. 17. As is seen, there is dark spot under the ball. The spots arise due to heating the plastic. Thus, the glowing penetrated through the water. The thickness of the water layer is about 1 cm. The video is presented in [Torchigin 2010] where GB is glowing on the bottom of the vessel filled by the water.

For the sake of justice, we need to note that we have observed many glowing hot balls properties of which coincide with that of the liquid drop heated at high temperature. These balls are crumble into small fragments in a collision with any obstacle. The color of these ball changes gradually from white to yellow, orange, red, dark red. The same takes place with any conventional body.

We can mark the following obvious anomalies

1. The brightness of the GB is not changed in time unlike the brightness of an incandescent body heated at very high temperature at which the body is cooling down gradually. In this case, the white color of the body becomes gradually yellow, orange, red, dark red and so on.

2. There is an abnormal great number of jumping from the surface of the fax paper. A conventional small ball made off any known material cannot jump on the paper.

3. There is smoky-colored trace after bouncing GB.

4. GB heats up the surface at a collision with it.

5. A number of jumps on the surface depends on properties of the surface. The maximal number corresponds to the surface that is heated maximally by the GB. This is a thin black leaf of paper. There are no jumps on the solid metal surface. A great GB that collides with the horizontal transparent windowpane of 4 mm thickness is crumpled into small fragments. A small GB stops near the region of colliding.

Conclusion

Analysis of the numerous experimental data obtained at attempts to produce a Ball Lightning in a laboratory in last two centuries shows that anomalous and unexplained phenomena can be explained on the assumption that there are incoherent optical spherical space solutions in nature. A physical nature of many so-called autonomous objects produced in laboratories falls into a wide circle of optical phenomena connected with self-confinement and self- organization of intense light in gaseous nonlinear optical mediums. Natural Ball Lightning is a particular case of these phenomena

Our analysis of experiments of Nauer in the middle of twentieth century testifies that the properties of the luminous object described by Nauer coincide completely with properties of light balls derived by us theoretically.

Our analysis of properties of autonomous objects arising in the field of the intense light in the vapor of liquid nitrogen enables us to discover the phenomenon of the self-organization of the intensive light propagating in a gas [Torchigin 2004 Opt. Comm.].

References

1996 Torchigin V.P. Amplification of light in lightguides and resonators formed by acoustic wave. J. Tech. Phys. 66(8) 1996 107.

1996 Torchigin V.P. Conversion of the light in a focon by the use of an acoustic wave as a pump. J. Tech. Phys. 66 (4), 128 (1996).

2002 Torchigin V.P. 2002. About stability of spherical layers of compressed air formed by intense light. Investigated in Russia. Electronic Journal. http://zhurnal.ape.relarn.ru/articles/2002/093.pdf (In Russian).

2003 Torchigin V. P. On the nature of Ball Lightning, Doclady Physics vol. 48, no. 3 pp. 108-11 (2003).

2003 Torchigin V. P., 2003 Optical Resonators in the Atmosphere. Laser Physics 13, no. 6, 919–931. 2003

2003 Torchigin V.P., A.V. Torchigin An increase in the wavelength of the light pulses propagating through a fiber. Physics Letters A, 311 (2003) 21.

2003 Torchigin, V. P. Lomonosov; 2003 no.2, 86-90.

2003 Torchigin, V. P., Torchigin, A. V. Chemistry and life, 2003, № 1, 12–15.

2003 Torchigin, V. P., Torchigin, A. V. Propagation of self-confined Light radiation in Inhomogeneous Air. PhysicaScripta, 2003, 68, 388–393.

2003. Torchigin V. P., Torchigin S. V., 2003 Optical solitons at propagation of whispering gallery waves. Quantum Electronics, 33 (10), (2003), 913–918.

2003. Torchigin V. P., V.A. Suchugov, I.K. Krasuyk et al., 2003 Change in the wavelength of light radiation stored within an optical resonator by means of an acoustic pulse. Optics

2004 Torchigin V. P.Acousto-optical devices USA patent number 6771412 of 3 August 2004.

2004 Torchigin V.P., Torchigin A.V. Behavior of self-confined layer of light radiation in the air atmosphere. Phys. Lett. A. 2004, 328/2–3, 189–195.

2004 Torchigin, V. P. Manifestation of Optical Quadratic Nonlinerity in Gas Mixtures. Physics; 2004, 49, No.10, 553–555

2004 Torchigin, V. P., Torchigin A. V., Space soliton in gas mixtures. Opt. Comm. 2004 240/4-6, 449-455

2004 Torchigin, V. P., Torchigin, A. V. Mechanism of the Appearance of Ball Lightning from Usual Lightning. Doclady Physics; 2004, 49, No. 9, 494–495

2004. Torchigin V. P., Torchigin A. V., 2004 Role of Ball Lightnings in Low Energy Nuclear Reactions. Infinite Energy 54, (2004), 46–50

2005 Torchigin V. P., A. V. Torchigin, 2005 Physical Nature of Ball lightning. European physical Journal D 36, (2005), 319–327.

2005 Torchigin V.P., A.V. Torchigin, Features of Ball Lightning stability, Europhysics Journal D 2005, 32, 383–389.

2005 Torchigin V.P., A.V. Torchigin, Phenomenon of ball Lightning and its outgrowth. Phys. Lett. A; 2005, 337, 112–120.

2006 Torchigin V.P. Is it possible to consider the Ball lightning as a reason of the Chernobyl tragedy? Bulletin of Atomic Energy 84 89-92.

2007 Torchigin, V. P., Torchigin, A. V. Self-organization of intense light within erosive gas discharge. Phys. Lett. A; 2007, 361, 167–172.

2010 V.P. Torchigin V.P., A.V. Torchigin On phenomenon of light radiation from miniature balls immersed in water, Physics Letters A 374 (2010) 588-591

2010 V.P. Torchigin, A.V. Torchigin Ball Lightning as an Optical Incoherent Space Spherical Soliton. In Handbook of Solitons: Research, Technology and

Applications. Editors S.P. Lang and Salim H. Bedore. Novapublishers (2010) 3-54

2011 V.P. Torchigin, A.V. Torchigin Chapter 6 Ball Lightning as an Optical Incoherent Space Spherical Soliton. In book Lightning: Properties, Formation and Types Editor Matthew D. Wood Novapublishers (2011) 133-184.

2012 Torchigin V.P., Torchigin A.V. Comparison of various approaches to the calculation of optically induced forces, Annals of Physics,Volume 327, Issue 9, September 2012, Pages 2288–2300 2012

2012 Torchigin V.P., Torchigin A.V. Interrelation between striction forces in dielectrics and optically induced forces in transparent media Physica Scripta Volume 86 Number 2 2012 86 025402

2013 Torchigin V.P., Torchigin A.V. Comment on ``Transverse radiation force in a tailored optical fiber'' Physical Review A 2013 Vol. 88 p. 027801 013Torchigin V.P., Torchigin A.V. Compensation of the optically induced Lorentz force in a homogeneous optical medium Optik2013 Vol.124, p.5492-5495

2013 Torchigin V.P., Torchigin A.V. Interrelation between Ball Lightning and optically induced forces. PhysicaScripta013 Vol.88 number 3, p. 035402 2013

2014 Torchigin V.P., Torchigin A.V. Optically induced force in a curve lightguide. EPJ AP European physical journal Applied Physics 2013, vol. 63, p. 10501

2014 Torchigin V.P., Torchigin A.V. Comment on "Theoretical analysis of the force on the end face of a nanofilament exerted by an outgoing light pulse" Physical Review A 2014, vol. 89, page 057801

2014 Torchigin V.P., Torchigin A.V. Magnitude of the photon momentum in matter. American Journal of Science and Technology 2014 Vol.4 Nom. 4 page 151-156

2014 Torchigin V.P., Torchigin A.V. Pressure Exerted on a Semi-Infinite Lossles Dispersionless Dielectric by a Plane Electromagnetic Wave OPEN JOURNAL OF MODERN PHYSICS 2014 Vol. nom. 3 1 page 1-7.

2014 Torchigin V.P., Torchigin A.V. Propagation of a light pulse inside matter in a context of the Abraham–Minkowski dilemma Optik 2014 vol. 125, issue 11, pp. 2687-2691.

2014 Torchigin V.P., Torchigin A.V. Resolution of the Age-Old Dilemma about a Magnitude of the Momentum of Light in Matter Physics Research International 2014, Vol. 2014, Pages 126436.

2014 Torchigin V.P., Torchigin A.V. The momentum of an electromagnetic wave inside a dielectric derived from the Snell refractive law Annals of Physics 2014, vol. 351, pages 444-446

Abrahamson, J.,Dinnis, J. Nature2000, 403, 3, February,519-521.

Avramenko R. F.Ed. Ball Lightning in a laboratory, (Moscow, Himiya, 1994).

Barry J. D., Ball Lightning and Bead Lightning, Plenum Press: NY, 1980.

Bychkov V. L. Bychkov A. V., Timofeev I. B J. Tech Fiz 2004, 74, issue 1, 128–133.

Carrel, Mike, (1996). In Infinite Energy Magazine Special Selection pp. 62–70.

Dmitriev M. T. Priroda 1967, 6, 98.

Egorov A. I., Stepanov S. I. J. Tech Fiz. 2002, 72, issue 12, 102–104.

Egorov A.I., Stepanov S.I., Shabanov G.D. UspehiFiz 2004 174(1), 107

Emelin S. E. et al (1994) Physical conditions of the ball lightning ejection caused byinteraction of electrical discharge with metal and polymer;http//www.balllightning.narod.ru/isbl01/BL_eject.htm

Emelin S. E. et al J. Tech Fiz, 1997, 67, № 3, 19–28.

Gezehus N. A. Journal of Russian chemical physical society, 1900, 8, 311.

Gooses V., P. de Graaf and R. Dekker, 2009
http://www.youtube.com/watch?v=QLTPELhKAYM and
http://www.youtube.com/watch?v=QLTPELhKAYM&mode=related&search

Grigoriev A. I., Grigorieva I.D., Shiryeva S.O. J. Scientific Explorations, 1992, 261–279.

Grigoriev I. S. Handbook of physical Quantities, Ed. (Energoatomizdat, Moscow 1991; CRC Press, Boca Raton, 1997)

Jones A. T. Science (New Series), 1910, 31, 144.

Klimov, A. I.; Mishin, G. I. Letters in J. Tech. Fiz. 1993. 18 (13), 19.

Klimov, A., I.; Malchenko, D. M.; Sukovatkin, N.,N. In Ball lightning in laboratory; Avramenko R. F.; Ed.; Himiya: Moscow, 1994.

Ledus, S. Compt. Rend.; 1899, 129, 37.

Lewis, E. H., In Tenth International Conference on Cold Fusion, USA, Massachusetts, Cambridge, August, 2003.

Luybimow, G. A.; Rahovskiy, V.,I.Usp. Fiz. Nauk; 1978, 125, 665.

Marum, Van. Phil. Mag;1800, 8, 313.

Mesyats, G. A. Ectons in vacuum discharge: discharge, spark, arc; Nauka: Moscow, 2000.

Mesyats, G.A., 1995. Uspehi Physics;1995 165 (6), 601–626.

Mishin, G. I, Klimov, A. I; Gridin, A. Yu. Letters in J. Tech. Fiz; 1992 18(6), 37-43.

Paiva, G. S. et al. Phys. Rev. Letters; 2007 PRL 98, 048501 January 2007.

Pirozerski, A. L., Emelin, S. E. (2003)
http://www.balllightning.narod.ru/golub02/mw/Slides/doklad_eng.html

Shoulders, K. Infinite Energy, 2005, 61.

Singer, S. The Nature of Ball Lightning; Plenum Press: NY, 1971.

Stakhanov I. P. The physical nature of Ball Lightning (Atomizdat, Moscow 1979 CEGB trans CE 8244)

Toepler, M. Ann. Phys.; 1901, IV 6, 339.

Urutskoev, L. I. Lomonosov; 2002, 10, 8-12 (In Russian).

Urutskoev, L. I., Fillipov D. N.. Usp. Fiz. Nauk; 2004, 174, № 12, 1355–1358.